Unveiling the Japan Earthquake

A Tale of Unity, Recovery, and Hope

Maria D. Johnson

INTRODUCTION

In the wake of a seismic event that unfolded on the 1st of January, the very ground beneath Japan trembled, leaving in its wake a tapestry of resilience, human spirit, and the unyielding force of nature. This eBook, "Unveiling the Japan Earthquake," seeks to narrate the unfolding events, capturing the essence of a nation grappling with the aftermath of a powerful earthquake.

As we turn our attention to the events on the 1st of January, we embark on a journey to understand the impact of this seismic force on the lives of the Japanese people, their communities, and the collective spirit that binds them together. With a lens focused on both the immediate consequences and the unfolding recovery efforts, this eBook endeavors to provide a comprehensive and compassionate account of a nation coming to terms with a sudden, unexpected shift in its landscape.

IMPACT OF THE EARTHQUAKE

Monday saw a strong earthquake hit central Japan that resulted in the destruction of buildings, the loss of electricity to tens of thousands of households, the death of one person, and the evacuation of some coastal residents to higher ground.

The 7.6-magnitude first earthquake caused waves to break off throughout Japan's west coast and into neighboring South Korea, measuring around one meter.

For the first time since the earthquake and tsunami that slammed northeast Japan in March 2011 and killed around 20,000 people, the Japan Meteorological Agency (JMA) first issued a major tsunami warning for the prefecture of Ishikawa. Later, it was decreased, and finally, it was reduced to an advisory.

The U.S. Geological Survey claims that the earthquake was the strongest to have struck the area in over 40 years.

Reporters were informed by government spokesman Yoshimasa Hayashi that houses were demolished, fires started, and army troops were sent to assist in rescue efforts.

According to broadcaster NTV, which cited local police, an old man was declared deceased when a structure collapsed in Shika Town, Ishikawa.

In the prefecture, pictures from local media showed a structure in Suzu crumbling in a cloud of dust and a massive fissure in a road in Wajima where parents grabbed their children in a state of terror.

A social media user shared a video of the Keta Grand Shrine in Hakui, close to the shore, trembling during the earthquake while a large number of onlookers looked on. "It's swaying," she lets out. "This is scary!"

To commemorate the beginning of the new year, millions of Japanese people visit shrines and temples on January 1.

Images captured the remains of a broken stone gate scattered at the entrance of another temple, with worried worshipers observing, in the adjacent tourist hotspot of Kanazawa.

The adjacent Nagano prefecture's mountains felt the tremor as well.

"Everyone was panicked because the snow from the electric wire (came down) and also from the roof it fell down and all the cars are shaking," Taiwanese traveler Jonny Wu, who is on a skiing vacation in Nagano, told Reuters.According to JMA official Toshihiro Shimoyama, there may be additional powerful earthquakes in the area in the next days. Seismic activity in the area has been simmering for more than three years.

Additionally, North Korea and Russia issued tsunami warnings for specific regions.

According to the Japanese government, as of Monday night, it has issued evacuation orders to nearly 97,000 residents of nine prefectures located on Honshu's western coast. Sports halls and school gymnasiums—which are frequently utilized as evacuation centers during emergencies—were where they were supposed to spend the night.

Ayako Daikai, a resident of Kanazawa, claimed she and her husband and two kids had fled to a neighboring elementary school shortly after the earthquake struck. She reported that there were a lot of evacuees in the gym, stairwells, halls, and classrooms.

When Reuters called, she responded, "We haven't decided when to return home yet."

NUCLEAR PLANTS

Even though access to quake-affected areas was difficult owing to blocked roads, Japanese Prime Minister Fumio Kishida told reporters late on Monday that he has ordered search and rescue personnel to do everything necessary to preserve lives.

Following the catastrophe, the Imperial Household Agency announced that it will postpone Emperor Naruhito and Empress Masako's scheduled New Year's appearance for Tuesday.

The earthquake strikes at a crucial moment for Japan's nuclear sector, which has been confronted with strong resistance from certain residents ever since the earthquake and tsunami that occurred in 2011 and led to the Fukushima nuclear meltdowns. The calamity destroyed entire communities.

The largest nuclear power station in the world, Kashiwazaki-Kariwa, was placed under an operating prohibition by Japan last week. The plant has been shut down since the 2011 tsunami.

According to the Nuclear Regulation Authority, no anomalies have been verified at any nuclear plants by the Sea of Japan, including the five operational reactors at the Ohi and Takahama plants in Fukui Prefecture owned by Kansai Electric Power (9503.T).

The nearest nuclear power plant to the epicenter, Hokuriku's Shika facility in Ishikawa, had already stopped operating its two reactors before the earthquake for routine inspections and had not experienced any seismic damage, according to the agency.

"TSUNAMI! EVACUATE!"

Television displays displayed a bright yellow message that said, "Tsunami! Evacuate!" after the earthquake, warning those living in certain coastal locations to leave right once.

The town of Wajima, which has 30,000 residents and is well-known for its lacquerware, was reported to have at least 30 collapsed structures, and many buildings were destroyed by fire.

A 500-kilometer distance from Wajima on the opposite shore, Tokyo's structures were also shaken by the earthquake.

With temperatures expected to plummet to around freezing tonight in certain regions, utilities supplier Hokuriku Electric Power (9505.T) said that about 32,000 homes in the prefecture of Ishikawa were still without power as of late on Monday.

In the neighboring prefecture of Niigata, 700 families still lack electricity, according to Tohoku Electric Power (9506.T).

Between Kanazawa and Toyama, there are four bullet train services that have halted. As of late Monday, West Japan Railway (9021.T) stated that 1,400 people were still stranded on these trains.

According to transport authorities, flaws in the runway prompted the closure of one of Ishikawa's airports.

While Japan Airlines (9201.T) canceled the majority of its flights to the Niigata and Ishikawa regions, ANA (9202.T) turned back aircraft bound for airports in Toyama and Ishikawa.

On Monday afternoon, a 7.5-magnitude earthquake slammed western Japan, resulting in the collapse of buildings, fires, and tsunami warnings that reached as far away as eastern Russia. The earthquake also caused citizens to be ordered to evacuate the impacted coastal districts of Japan.

The United States Geological Survey (USGS) reports that the earthquake occurred on the Noto Peninsula in Ishikawa prefecture around 4:10 p.m. local time, at a depth of 10 kilometers (6 miles).

Just over ten minutes after the Japan Meteorological Agency promptly issued a tsunami warning for western Japan's coastal districts, the first waves were reported to have impacted the coast.

Early reports, according to Japanese public broadcaster NHK, came from the city of Wajima in the Ishikawa prefecture, which saw tsunami waves of around 1.2 meters (3.9 feet) at 4:21 p.m. There were no recorded immediate damages. Later, the tsunami warnings were lifted for a stretch of Japan's western coast.

Defense Minister Minoru Kihara told reporters earlier on Monday that the defense ministry has sent out 1,000 military men to aid in the rescue and recovery work.

There have been reports of injuries as well as damage to structures. According to NHK, the city's police reported that several individuals were stranded in damaged homes. As of yet, no fatalities have been recorded.

According to Suzu hospital authorities, they received injured patients. They also mentioned that several physicians were unable to report to work due to damaged roads, as NHK reported.

For the first time since the terrible earthquake in 2011, the Japan Meteorological Agency issued a "major tsunami warning" for Noto, although it was later lowered to a "tsunami warning."

Later, when the likelihood of more tsunami waves decreased, tsunami advisories were lifted, while warnings for waves up to one meter (3 feet) in height remained in effect.

Waves predicted to be less than one meter are classified as "tsunami advisory" under Japan's tsunami warning system, waves predicted to be up to three meters as "tsunami warning," and waves predicted to be more than five meters as "major tsunami warning."

Japan's Chief Cabinet Secretary Yoshimasa Hayashi encouraged residents in regions under tsunami warnings to evacuate to higher ground earlier in a televised speech.

'The whole room was shaking'

When the earthquake struck Ishikawa prefecture, NHK footage showed cameras trembling violently as waves crashed into the beach.

The earthquake also shook homes; pictures reveal fallen roofs and trembling foundations.

when he waited for a bus home in western Japan, an eyewitness described seeing people "panicked" when the earth began to shake.

You can see that all of the snow from the roof and the electric line has fallen, and all of the automobiles are trembling. And so at that moment, everyone was in a panic," Taiwanese traveler Johnny Wu told Reuters.

When the earthquake struck, Baldwin Chia, a 38-year-old Shanghai visitor who was with a group of snowboarders, told Reuters that "the whole room was shaking, the TV was shaking." Everything has to stay on the table. Nonetheless, I felt secure in my chamber. Everything else, though, was trembling.

In the Ishikawa prefecture of Japan, four trains between the cities of Toyama and Kanazawa were halted, and some Shinkansen bullet train services were suspended. More than a decade after the earthquakes started, about 1,400 people are still stuck inside the stationary high-speed trains, according to a report from NHK that referenced Japan Railways West.

Videos posted on social media captured the quake's aftermath, with merchandise scattered throughout store aisles. In one video shot from inside a train, the tremor caused the signposts on the platform to move violently.

After the earthquake, more over 32,500 residences in Ishikawa prefecture lost electricity, according to Hokuriku Electric Power Company.

In a statement on X, Japan's Kansai Electric Power Company stated that no anomalies had been recorded at any of the region's nuclear reactors.

Chief Cabinet Secretary Yoshimasa said that there had been "no major results" from the impacted power converter at the Shika nuclear power station in Ishikawa. According to Japan's Nuclear Regulation Authority, the power plant's reactors are operating without issue, NHK said.

The USGS reported that many strong aftershocks occurred shortly after the major earthquake.

Around 4 kilometers (2.4 miles) southwest of Anamizu, at 4:18 p.m. local time, an aftershock with a magnitude of 6.2 hit at a depth of 10 kilometers (6 miles), according to the USGS.

The USGS reports that a further 5.6 magnitude aftershock was detected closer to the main quake, while 5.2 magnitude vibrations were registered some 58 kilometers (about 36 miles) distant.

The nation's meteorological department issued a warning, alerting people to the possibility of building collapses and landslides, and stating that strong aftershocks may last for up to a week.

Defense Minister Kihara informed reporters that 1,000 military soldiers had been sent by the ministry to assist in the rescue and recovery operations.

According to Japanese Prime Minister Fumio Kishida, officials are assessing possible damage in impacted areas.

"The Prime Minister's Office of Response - Disaster CounterMeasure Headquarters has been established immediately. The prime minister posted on X, the previous Twitter platform, on Monday. "We are putting forth all efforts in disaster response, putting human lives as priority and making every effort to assess damages."

A strong earthquake that caused waves to reach over a meter in height and prompted tsunami advisories and evacuation orders was felt throughout the western shore of central Japan.

At around 4.10 p.m. local time (07.10 GMT), an earthquake with an estimated magnitude of 7.6 hit the Noto peninsula in Ishikawa prefecture on the main central island of Honshu. Tens of thousands of houses lost electricity as a result, and train and aviation services were also affected.

Four deaths were verified early on Tuesday by Ishikawa prefecture officials, as reported by the Kyodo news agency. Reports surfaced of persons buried beneath the fallen home rubble in Wajima city, near the epicenter. People in the city were evacuated in the dark as a massive fire consumed a row of homes. Numerous injured individuals were reportedly arriving at local hospitals for treatment, despite the fact that patient transportation was hampered by damaged roadways.

According to the Japan Meteorological Agency (JMA), the area was struck by many earthquakes with a minimum magnitude of 3.5 between 4 and 7.30 p.m. At first, the JMA issued lower-level tsunami warnings or advisories for the remainder of Honshu and Hokkaido, the northernmost of Japan's main islands, in addition to a significant tsunami warning for Ishikawa. A few hours later, the warning was reduced to a standard tsunami, indicating the seas may still rise to three meters (10 feet).

About four hours after the 7.6-magnitude earthquake, the US Pacific Tsunami Warning Center declared that the tsunami risk had "largely passed." The center had issued a warning about potential tsunamis along 190 miles of shoreline.

Along a damaged road in Wajima, Ishikawa prefecture, people stroll. Image courtesy of Kyodo/Reuters

According to Japanese Prime Minister Fumio Kishida, authorities were still determining the full extent of the devastation and citizens needed to brace themselves for additional earthquakes.

He stated, "I urge people in areas where tsunamis are expected to evacuate as soon as possible and residents need to stay on alert for further possible earthquakes." In Tokyo, buildings could be felt 190 miles (300 km) distant.

The NHK broadcaster had warned viewers
to seek higher ground as soon as possible
and to be on the lookout for more tsunamis
and earthquakes.

The person stated, "We're not that close to
the coast so tsunamis are not a worry but
the aftershocks just keep coming and
shaking the whole house, it's terrifying." The
Tokyo resident was spending the new year
break in her birthplace of Takaoka, a city
just south of the worst-affected Noto
peninsula.

According to Hayashi Yoshimasa, a
spokesman for the government, citizens
should get ready for any potential

aftershocks and the Japan Self-Defense Forces are ready to be sent into action to assist with rescue and recovery operations.

The site of the earthquake in Ishikawa Prefecture, western central Japan, is seen on a map provided by the US Geological Survey.Image courtesy of USGS/EPA

NHK released footage that seemed to show Ishikawa buildings crumbling. In Ishikawa and Toyama prefectures, more than 36,000 households lost electricity, according to Hokuriku Electric Power, the utilities company.

While the telecom carriers SoftBank and KDDI reported phone and internet service outages in Ishikawa and Niigata prefectures, respectively, on their websites, high-speed rail services to Ishikawa have been discontinued.

While Japan Airlines canceled the majority of its flights to the Niigata and Ishikawa regions, the Japanese carrier ANA turned back aircraft bound for airports in Toyama and Ishikawa. One of Ishikawa's airports, according to the authorities, is closed.

Japan regularly conducts emergency exercises to get ready for a large shock and has tight construction laws designed to guarantee structures can survive powerful earthquakes.

More quickly than those on the Pacific coast, tsunamis from earthquakes off the coast of the Sea of Japan been known to reach land in less than ten minutes. It took the tsunami from the 2011 magnitude 9 earthquake around thirty minutes to reach the coast.

According to Japan's Nuclear Regulation Authority, no abnormalities have been detected at any nuclear power facilities near the Sea of Japan. This includes the five operational reactors at the Ohi and Takahama plants in Fukui prefecture, which are owned by Kansai Electric Power. The impacted coastline is home to six sites with a total of 22 reactors.

The agency stated that Hokuriku's Shika facility in Ishikawa, which was the closest to the epicenter of the earthquake, had already stopped its two reactors for a routine check before to the earthquake and was unaffected by it.

Japan is among the countries on Earth most vulnerable to earthquakes. On March 11, 2011, a massive earthquake and tsunami that hit the country's northeast caused the deaths of 18,000 people. Towns were completely destroyed, and the Fukushima Daiichi nuclear reactor had meltdowns due to the calamity. Since the accident, almost all of Japan's nuclear power facilities have been placed on hold.

This report was supported by Agence France-Presse and Reuters. People don't chose to support the Guardian for a good reason.

Right now, not everyone has the money to pay for the news. We decide to publish our work freely for everyone to read, including those in Nigeria, for this reason. If this describes you, feel free to read on without charge.

However, there are three solid reasons for you to decide to assist us right now if you are able to.

1. At a time when the wealthy and powerful are getting away with more and more, our superior investigative journalism serves as a scrutinizing force.

2. Because we are self-governing and don't have a wealthy owner in charge, your contributions directly fund our reporting.

3. It requires less time and doesn't cost as much as reading this letter.

Whether you can contribute a little or a lot
more, decide to support the Guardian's work
for years to come.

Thousands of people in Japan are spending the night in evacuation centres after a powerful earthquake.

According to the Kyodo news agency, four deaths have been verified, and several injuries have been reported.
In many places, an undetermined quantity of individuals remain buried beneath the debris of crumbling structures.
On Monday, at around 16:10 local time (07:10 GMT), a 7.6-magnitude earthquake occurred. After being issued, tsunami warnings were eventually lowered.
Approximately sixty earthquakes have been detected since the first one.

In the Hakuba Alps of Japan, a snowboarder on vacation reported that his hotel room shook. Baldwin Chia told Reuters that while he was worried about avalanches, he had not heard of any occurring.

Although earthquakes are frequently reported in Japan, he remarked that "you wouldn't expect to actually experience one". When the earthquake struck, British visitor to Japan Andy Clark told the BBC that it was a "scary afternoon and evening" because he was in the seaside city of Toyama.

According to him, he "grabbed the sea wall to stay upright" before running for cover on the roof of a school. Mr. Clark claimed that the aftershocks were making it "hard to get some sleep".

Professor Jeffrey Hall of Kanda University reported experiencing earthquakes for around two minutes while in Yokohama, which is located on the other side of Japan's main island. He described the earthquake as "very, very serious".

Towns and communities in central Japan were obliged to relocate their residents into shelters for the night.

Although significant infrastructure damage is obvious, the entire scale of the disaster is probably not going to be known until Tuesday morning.

National network NHK reported that numerous houses and power poles toppled, according to officials in Suzu City, Ishikawa prefecture.

According to utilities company Hokuriku Electric electricity, over 36,000 households were left without electricity and major routes were closed close to the epicentre of the earthquake.

Rupert Wingfield-Hayes, a former BBC journalist in Japan, reported from Taiwan that a landslide had destroyed several hundred meters of the main freeway connecting Toyama and Kanazawa.

Images from Uchinada, in the prefecture of Ishikawa, revealed a road's surface to be fractured and rippled. There were also photos showing damage to the Onohiyoshi Shrine in Kanazawa.

Caption for video, watch: Kanazawa was shaken by an earthquake, causing alarm.

At first, officials in the coastal Noto district of Ishikawa, close to the epicentre of the quake, issued a significant tsunami warning, stating that waves may rise as high as 5 meters (16 feet).

According to local accounts, this was Japan's first warning of this kind since 2011, when a strong earthquake in the northeast caused waves as high as 40 meters.

Less than a meter high waves really crashed against the Sea of Japan beach in Ishikawa on Monday.

According to NHK, the significant alert was eventually reduced to a mere warning, and finally to an "advisory". The neighboring prefectures of Niigata and Toyama were also on guard.

Because of its location on the so-called Pacific Ring of Fire, where many tectonic plates converge, Japan is among the seismically active countries on Earth. Japan has one of the most advanced tsunami warning systems in the world because of the ongoing threat of earthquakes.

Although there are many nuclear power plants in the impacted areas, there is "no risk of radioactivity leaking" from the facilities, according to Japan's nuclear authorities.

Following the earthquake, Russia and South Korea's meteorological office also issued tsunami warnings.

Following a 9.0 magnitude earthquake in 2011, a tsunami ripped across coastal areas in northeastern Japan, killing almost 18,000 people and uprooting tens of thousands more.

The most catastrophic nuclear catastrophe since Chernobyl was brought on by those tsunami waves, which caused a nuclear meltdown at the Fukushima power facility.

HISTORY OF EARTHQUAKE IN JAPAN

Discrete slow fault slides cause slow earthquakes. They, together with rapid, frequent earthquakes, are an essential part of intraplate deformation processes. The subduction zone along which the 9.0 magnitude Tohoku-Oki earthquake occurred on March 11, 2011, is known as the Japan Trench. Recent seismological and geodetic investigations have shown detailed slow seismic activity throughout this zone. This study reviews the history of research on slow earthquakes along the Japan Trench, including observational, experimental, and simulation studies.

We present an integrated slow earthquake distribution along the Japan Trench by combining the observations of related fault slip phenomena (e.g., small repeating earthquakes, earthquake swarms, and foreshocks of large interplate earthquakes) with slow earthquake observations (e.g., tectonic tremors, very-low-frequency earthquakes, and slow slip events). Megathrust and slow earthquakes are complementary in their geographical distribution, and slow earthquakes can occasionally set off nearby quick earthquakes. The massive interplate locked zone of the middle Japan Trench correlates with an approximately 200-km-long along-strike gap of seismic slow earthquakes (i.e., tectonic tremors and very-low frequency earthquakes).

The 9.0 Mw This locked zone was broken by the Tohoku-Oki earthquake, however the break did not extend very far into the slow-earthquake-genic areas of the northern and southern Japan Trench. In the Japan Trench, slow earthquakes have a role in both the rupture initiation and termination stages of megathrust earthquakes. We then described the geological environment of the slow-earthquake-genic regions, including water sources, pressure–temperature conditions, and metamorphism, and compared the integrated slow earthquake distribution with the crustal structure of the Japan Trench (e.g., interplate sedimentary units, subducting seamounts, petit-spot volcanoes, horst and graben structures, residual gravity, seismic velocity structure, and plate boundary reflection intensity). We were able to have a thorough discussion on the part that slow earthquakes played in the Tohoku-Oki earthquake's occurrence process because to the integrated slow earthquake distribution.

The correlation between the distribution of slow earthquakes and the crustal structure and geological environment offers valuable insights into the formation of slow earthquakes in the Japan Trench. It also suggests that a comprehensive knowledge of the complicated slow earthquake distribution requires an understanding of highly overpressured fluids. Moreover, we suggest that thorough monitoring of slow earthquake activity might enhance intraplate seismicity projections in the Japan Trench.

INTRODUCTION OF HISTORY

A subduction zone called the Japan Trench is situated off the eastern Pacific coast of Japan.
The Pacific Plate is subducting beneath the Okhotsk Plate in this trench at a pace of around 9 cm/yr (Bird 2003), with an estimated age of 130–140 Ma (Müller et al. 2008).

In this subduction zone, several large-scale interplate earthquakes have been detected and are being intensively explored. Moment magnitude (Mw) 7–8 megathrust earthquakes have often occurred at a depth of 20–50 km in the central (37–39° N) and northern (39–41° N) Japan Trench (Nagai et al. 2001; Murotani et al. 2003; Yamanaka and Kikuchi 2002).

On March 11, 2011, the Mw 9.0 Tohoku-Oki earthquake shattered the central and southern regions of the Japan Trench, and it was the strongest earthquake ever recorded in Japan.

The Tohoku-Oki rupture reached the trench axis in the central Japan Trench (37–39° N), and the Japan Trench megathrust slipped more than 50 m (e.g., Lay 2018; Kodaira et al. 2020, 2021), as demonstrated by bathymetric changes (Fujiwara et al. 2011), crustal deformation (Iinuma et al. 2012), and seismic waveforms (Ide et al. 2011).

Small interplate earthquakes in the Japan Trench have also been extensively investigated, in addition to megathrust earthquakes. Small repeating earthquakes have been reported in the Japan Trench

(Matsuzawa et al. 2002; Igarashi et al. 2003; Uchida and Matsuzawa 2013). These earthquakes cause comparable ruptures to occur repeatedly in the same place on the plate boundary (e.g. Nadeau and Johnson 1998).

A great deal of research has been done on the regularity and diversity of their rupture patterns (e.g Okada et al. 2003 Uchida et al. 2007; Okuda and Ide 2018; Ide 2019; Chang and Ide 2021).

The activity of slow earthquakes, another significant category of seismic phenomena, had remained poorly resolved until recently, despite the fact that much has been discovered about fast, frequent earthquakes in the Japan Trench. According to Ide et al. (2007), the phrase "slow earthquakes" refers to a broad range of fault sliding occurrences that last far longer than typical, quick earthquakes with the same seismic moment. Subduction zones like the Nankai Trough, which is situated off the Pacific coast of southwest Japan, are actively studying slow earthquakes, which have been seen in other circum-Pacific subduction zones.

The complicated distribution and occurrence patterns of several slow earthquakes were discovered in the Nankai Trough, in contrast to the Japan Trench (e.g., Obara 2002; Ito et al. 2007; Obara and Kato 2016).

In addition, it is interesting that in the Nankai Trough, the young and warm Philippine Sea Plate (about 15–25 Ma) is subducting (Müller et al. 2008), in contrast to the Japan Trench, where the ancient and cold Pacific Plate is subducting.

According to some research (e.g., Peacock and Wang 1999; Katayama et al. 2012), this has led to notable disparities in subduction zone processes (e.g., slow and fast earthquakes, arc volcanism, and metamorphism) between these two subduction zones.

Since the Tohoku-Oki earthquake in 2011, there has been an acceleration in slow earthquake research in the Japan Trench. A Mw 7.0 short-term slow slip event (SSE) accompanied by an earthquake swarm was reported around a month before the Tohoku-Oki earthquake on the shallow plate contact of the central Japan Trench (Kato et al. 2012; Ito et al. 2013; a green rectangle inside the Tohoku-Oki earthquake rupture). Additionally, while systematic and thorough detection of slow earthquakes along the entire Japan Trench has not been achieved, detection based on a visual inspection of broadband seismograms (Matsuzawa et al. 2015) and an analysis of repeating earthquakes (Uchida et al. 2016) also provided important insights into the activity of slow earthquakes (very-low-frequency earthquakes and SSEs, respectively) in the Japan Trench.

The detailed distribution of tectonic tremors, a type of slow earthquake, along the entire Japan Trench was finally revealed in 2019 thanks to pop-up-type ocean-bottom seismometers installed in the southern part of the Japan Trench (Ohta et al. 2019) and the Seafloor Observation Network for Earthquakes and Tsunamis along the Japan Trench (S-net) (NIED 2019). Using onshore seismic and geodetic observation networks, the systematic and thorough identification of other slow earthquake types (very-low-frequency earthquakes and short-term SSEs) followed the tremor discovery (Baba et al. 2020; Nishimura 2021).

The slowly occurring earthquake distribution throughout the Japan Trench is starting to become fully apparent as a result of these mounting study findings.

We examine the studies on slow earthquakes along the Japan Trench and provide an overview of their study history, starting from the first report of a substantial transient aseismic slip in the Japan Trench (Kawasaki et al. 1995) and continuing up to the present. Our analysis is based on the above research advancements.

We present an integrated slow earthquake distribution along the Japan Trench by combining the observations of related fault slip phenomena (e.g., small repeating earthquakes, earthquake swarms, and foreshocks of large interplate earthquakes) with slow earthquake observations (e.g., tectonic tremors, very-low-frequency earthquakes, and slow slip events). We address the occurrence mechanisms of megathrust earthquakes in the Japan Trench, focusing on the function of slow earthquakes, based on the distribution of slow earthquakes.

Additionally, we investigate the circumstances leading to slow earthquakes by contrasting the distribution of slow earthquakes with the crustal structure and geological setting of the Japan Trench.
We also note gradual seismic activity near the southern end of the Kuril Trench (41–43° N), in addition to the Japan Trench (34–41° N). But we confine our primary conversation to the Japan Trench. The areas

between 39 and 41° N (Iwate-Aomori-Oki), 37 and 39° N (Fukushima-Miyagi-Oki), and 34 and 37° N (Boso-Ibaraki-Oki) are referred to as the northern, central, and southern Japan Trenches, respectively, from this point on.

Readers interested in the 2011 Tohoku-Oki earthquake and related events instead of slow earthquakes are directed to other review articles (Lay 2018; Kodaira et al. 2020, 2021; Uchida and Bürgmann 2021). This is because the focus of this review study is on slow earthquakes along the Japan Trench.

A succinct overview of slow earthquakes

A brief introduction to slow earthquakes at subduction zone plate boundaries is given in this part, with an emphasis on subjects that are crucial to the sections that follow.

Peng and Gomberg (2010), Beroza and Ide (2011), Ide (2014), Saffer and Wallace (2015), Obara and Kato (2016), Bürgmann (2018), Obara 2020, Behr and Bürgmann 2021, Kirkpatrick et al.
2021 are only a few of the notable review papers on slow earthquakes that have been published. For further information, we direct readers to these papers. The exact occurrence patterns of some slow earthquakes were thoroughly studied by Obara (2020); these are not covered in this section. A thorough analysis of the connection between slow earthquakes and

megathrust earthquakes was given by Obara and Kato (2016). Three studies have provided detailed descriptions of the geological context and potential physical processes of slow earthquakes: Bürgmann (2018), Behr and Bürgmann (2021), and Kirkpatrick et al. (2021). Saffer and Wallace (2015) provide information on slow earthquakes on shallow subduction plate contacts.

A review on slow earthquakes in the Japan Trench

Discrete slow fault slides cause slow earthquakes. They, together with rapid, frequent earthquakes, are an essential part of intraplate deformation processes. The subduction zone along which the 9.0 magnitude Tohoku-Oki earthquake occurred on March 11, 2011, is known as the Japan Trench. Recent seismological and geodetic investigations have shown detailed slow seismic activity throughout this zone. This study reviews the history of research on slow earthquakes along the Japan Trench, including observational, experimental, and simulation studies. We present an integrated slow earthquake distribution along the Japan Trench by combining the observations of related fault slip phenomena (e.g., small repeating earthquakes, earthquake swarms, and foreshocks of large interplate earthquakes) with slow earthquake observations (e.g., tectonic tremors, very-low-frequency earthquakes, and slow slip events). Megathrust and slow earthquakes are complementary in their geographical distribution, and slow earthquakes can occasionally set off nearby quick earthquakes. The massive interplate locked zone of the middle Japan Trench correlates with an approximately 200-km-long along-strike gap of seismic slow earthquakes (i.e., tectonic tremors and very-low frequency earthquakes). This locked zone was broken by the Mw 9.0 Tohoku-Oki earthquake, however the break did not extend very far into the slow-earthquake-genic areas of the northern and southern Japan Trench. In the Japan Trench, slow earthquakes have a role in both the rupture initiation and termination stages of megathrust earthquakes. We then described the geological environment of the slow-earthquake-genic regions, including water sources, pressure–temperature conditions, and metamorphism, and compared the integrated slow earthquake distribution with the crustal structure of the Japan Trench (e.g., interplate sedimentary units, subducting seamounts, petit-spot volcanoes, horst and graben structures, residual gravity, seismic velocity structure, and plate boundary reflection intensity). We were able to have a thorough discussion on the part that slow earthquakes played in the Tohoku-Oki earthquake's occurrence process because of the integrated slow earthquake distribution. The correlation between the distribution of slow earthquakes and the crustal structure and geological environment offers valuable insights into the formation of slow earthquakes in the Japan Trench. It also suggests that a comprehensive knowledge of the complicated slow earthquake distribution requires an understanding of highly overpressured fluids. Moreover, we suggest that thorough monitoring of slow earthquake activity might enhance intraplate seismicity projections in the Japan Trench.

Introduction

Slow earthquakes are caused by discrete slow fault movements. Along with frequent and swift earthquakes, they constitute a crucial component of intraplate deformation processes. The Japan Trench is the name of the subduction zone that was affected by the 9.0 magnitude Tohoku-Oki earthquake that struck on March 11, 2011. Current geodetic and seismological studies have revealed extensive, sluggish seismic activity in this zone. The history of observational, experimental, and modeling investigations on slow earthquakes along the Japan Trench is reviewed in this paper. By combining the observations of related fault slip phenomena (e.g., small repeating earthquakes, earthquake swarms, and foreshocks of large interplate earthquakes) with the observations of slow earthquakes (e.g., tectonic tremors, very-low-frequency earthquakes, and slow slip events), we present an integrated distribution of slow earthquakes along the Japan Trench. The geographical distribution of megathrust and slow earthquakes complements one another, and slow earthquakes can occasionally trigger neighboring fast earthquakes. An approximately 200-km-long along-strike gap of seismic slow earthquakes (tectonic tremors and very-low frequency earthquakes) is correlated with the huge interplate locked zone of the middle Japan Trench. The Mw 9.0 Tohoku-Oki earthquake shattered this locked zone, however it did not go very far into the slow-earthquake-prone northern and southern Japan Trench. Both the rupture initiation and termination phases of megathrust earthquakes in the Japan Trench are influenced by slow earthquakes. After that, we discussed the water sources, pressure-temperature conditions, and metamorphism of the geological environment in the slow-earthquake-genic regions. Finally, we compared the integrated slow earthquake distribution with the Japan Trench's crustal structure, which includes interplate sedimentary units, subducting seamounts, petit-spot volcanoes, horst and graben structures, residual gravity, seismic velocity structure, and plate boundary reflection intensity. Thanks to the integrated slow earthquake distribution, we were able to have a detailed conversation about the role that slow earthquakes had in the Tohoku-Oki earthquake's occurrence process. Important information about how slow earthquakes develop in the Japan

Trench may be gleaned from the relationship between the distribution of slow earthquakes and the crustal structure and geological environment. It also implies that a thorough comprehension of the intricate distribution of slow earthquakes necessitates a working knowledge of severely overpressured fluids. Furthermore, we propose that careful observation of slow earthquake activity might improve intraplate seismicity estimates in the Japan Trench.

Types of slow earthquakes

There are five major types of slow earthquakes: low-frequency earthquakes (LFEs), tectonic tremors, very-low-frequency earthquakes (VLFEs), short-term SSEs, and long-term SSEs

Slow earthquakes are caused by discrete slow fault movements. Along with frequent and swift earthquakes, they constitute a crucial component of intraplate deformation processes. The Japan Trench is the name of the subduction zone that was affected by the 9.0 magnitude Tohoku-Oki earthquake that struck on March 11, 2011. Current geodetic and seismological studies have revealed extensive, sluggish seismic activity in this zone. The history of observational, experimental, and modeling investigations on slow earthquakes along the Japan Trench is reviewed in this paper. By combining the observations of related fault slip phenomena (e.g., small repeating earthquakes, earthquake swarms, and foreshocks of large interplate earthquakes) with the observations of slow earthquakes (e.g., tectonic tremors, very-low-frequency earthquakes, and slow slip events), we present an integrated distribution of slow earthquakes along the Japan Trench. The geographical distribution of megathrust and slow earthquakes complements one another, and slow earthquakes can occasionally trigger neighboring fast earthquakes. An approximately 200-km-long along-strike gap of seismic slow earthquakes (tectonic tremors and very-low frequency earthquakes) is correlated with the huge interplate locked zone of the middle Japan

Trench. The Mw 9.0 Tohoku-Oki earthquake shattered this locked zone, however it did not go very far into the slow-earthquake-prone northern and southern Japan Trench. Both the rupture initiation and termination phases of megathrust earthquakes in the Japan Trench are influenced by slow earthquakes. After that, we discussed the water sources, pressure-temperature conditions, and metamorphism of the geological environment in the slow-earthquake-genic regions. Finally, we compared the integrated slow earthquake distribution with the Japan Trench's crustal structure, which includes interplate sedimentary units, subducting seamounts, petit-spot volcanoes, horst and graben structures, residual gravity, seismic velocity structure, and plate boundary reflection intensity. Thanks to the integrated slow earthquake distribution, we were able to have a detailed conversation about the role that slow earthquakes had in the Tohoku-Oki earthquake's occurrence process. Important information about how slow earthquakes develop in the Japan Trench may be gleaned from the relationship between the distribution of slow earthquakes and the crustal structure and geological environment. It also implies that a thorough comprehension of the intricate distribution of slow earthquakes necessitates a working knowledge of severely overpressured fluids. Furthermore, we propose that careful observation of slow earthquake activity might improve intraplate seismicity estimates in the Japan Trench.

Relationship between slow earthquakes

Simultaneous occurrence of different types of slow and fast earthquakes

In close proximity, different kinds of slow earthquakes frequently happen concurrently. First identified in the deeper plate boundary of the Cascadia subduction zone, the coincidence of tectonic tremor bursts with short-term SSEs is known as episodic tremor and slip (ETS) (Rogers and Dragert 2003). Additionally, ETSs were found in the Nankai Trough in 2004 (Obara et al. 2004; Fig. 2d). Afterwards, it was discovered that VLFEs also happened during ETS (Ito et al. 2007; Ghosh et al. 2015). Moreover, it is known that long-term SSEs on the Nankai Trough's downdip side generate ETSs on the deeper plate interface, which is located at a depth of around 25 km (Hirose and Obara 2005). Many attempts have been made to estimate the source properties (e.g., source duration, rupture area, and moment) of short-term SSEs too small to be geodetically detected (Mw 5.5 or less) from characteristics of tremor bursts (e.g., Aguiar et al. 2009; Obara 2010; Wech et al. 2010; Gomberg et al. 2016). This is due to the remarkable correspondence between the short-term SSEs and tectonic tremors at deeper parts of the Cascadia subduction zone and Nankai Trough.

ETSs are not exclusive to the deeper subduction plate contact. Ocean-bottom seismometers and subseafloor borehole pore-fluid pressure measurements on the shallow plate interface (shallower than 10 km depth) of the Nankai Trough observed recurring ETSs (Araki et al. 2017). However, shallow ETSs have not been observed in the Cascadia subduction zone (McGuire et al. 2018).

Apart from the Nankai Trough and the Cascadia subduction zone, distinct ETSs have been detected in the subduction zones of Alaska (Rousset et al. 2019), Mexico (Rousset et al. 2017), and Costa Rica (Walter et al. 2013). But take note that quake bursts and short-term SSEs don't necessarily happen at the same moment. Ten days or more have elapsed between the occurrence of a short-term SSE and a seismic explosion in the Hikurangi Trench, a subduction zone off the east coast of New Zealand's North Island (Todd et al. 2018; Shaddox and Schwartz 2019; Nishikawa et al. 2021).

Moreover, certain short-lived seismic events in the Hikurangi Trench were devoid of tectonic vibrations that could be felt (Todd and Schwartz 2016). We cannot, however, completely rule out the possibility that there were tectonic shocks with magnitudes lower than the observed limit. This also applies to subduction zones in which no reports of a distinct ETS have been made.

Simultaneous slow and quick earthquakes are also possible.

Off the Boso Peninsula in the Sagami Trough in eastern Japan, there has been recurrent observation of the conjunction of short-term SSEs with swarms of interplate earthquakes (Mw 5) (Ozawa et al. 2003; Sagiya 2004). Recurring earthquakes were among these seismic swarms (Kato et al. 2014; Gardonio et al. 2018). These earthquake swarms are thought to be initiated by stress loading brought on by the SSE (Fukuda 2018). Similar SSE-fast earthquake coincidences have been noted in the subduction zones of Mexico (Liu et al. 2007), Ecuador (Vallée et al. 2013), Peru (Villegas-Lanza et al. 2016), the Japan Trench (Ito et al. 2013), and northern Chile (Socquet et al. 2017). However, in the Ecuador and Hikurangi trenches, swarms of intraslab or upper-plate earthquakes accompany short-term SSEs on the plate interface (Collot et al. 2017; Shaddox and Schwartz 2019; Nishikawa et al. 2021). Fast earthquakes that occur concurrently with SSEs are typically interplate earthquakes. Megathrust earthquakes, namely the 2011 Mw 9.0 Tohoku-Oki (Kato et al. 2012; Ito et al. 2013), 2014 Mw 8.1 Iquique (Ruiz et al. 2014; Socquet et al. 2017), and 2014 Mw 7.3

Papanoa (Radiguet et al. 2016) earthquakes, respectively, were the result of either long-term or short-term SSEs in the trenches of Japan, northern Chile, and Mexico. Numerous foreshocks coincided with the preseismic SSEs of the Tohoku-Oki and Iquique earthquakes. It is believed that these preseismic SSEs either initiated the megathrust earthquakes or had a role in their nucleation.

There are other types of slow earthquakes that can happen at the same time as quick earthquakes, besides SSEs. Coinciding tectonic tremors and swarms of interplate microearthquakes have been detected in the shallow section of the southern Japan Trench (Obana et al. 2021). Tectonic tremors, short-term SSEs, and swarms of intraslab earthquakes all happen at the same time in the middle Hikurangi Trench (Romanet and Ide 2019; Nishikawa et al. 2021). In addition, a recent research (Yamamoto et al. 2022) on the shallow plate contact of the Nankai Trough off the southeast coast of the Kii Peninsula documented the simultaneous occurrence of an SSE, VLFEs, and a swarm of interplate microearthquakes.

Unified understanding of slow earthquakes

Although slow earthquakes are diverse, efforts have been made to understand them in a unified manner. Furthermore, Ide et al. (2007) found that the seismic moment M_0 of slow earthquakes (LFEs, VLFEs, short-term

SSEs, and long-term SSEs) is approximately proportional to their source duration T (i.e., M_o T). In contrast, the seismic moment of fast earthquakes is proportional to the cubed source duration (i.e., M_o T^3), and there is a significant gap between the scaling relations of slow and fast earthquakes. Aseismic transients that fall into this gap have been reported in the northern Japan Trench (Kawasaki et al. 1995) and Izu–Bonin subduction zone (Fukao et al. 2021), but are very rare.

Ide et al. (2007) regarded multiple slow earthquakes as the same physical phenomenon with varying seismic moments based on the moment–duration scaling of slow earthquakes (Mo T). They thought that diffusion may be the cause of slow earthquakes, but it was unclear what physical mechanism—fluid diffusion, stress diffusion, or something else entirely—was at play. The parabolic migration patterns of deep tectonic tremors and LFEs in the Nankai Trough on a scale of several tens of kilometers (e.g., Ide 2010; Kato and Nakagawa 2020) demonstrate the diffusional aspect of slow earthquakes, while on a larger spatial scale (100 km or larger), deep tectonic tremors in the Nankai Trough and Cascadia subduction zone show almost constant-velocity (approximately 10 km/day) migration patterns (e.g., Ito et al. 2007; Houston et al. 2011; Nakata et al. 2011). As is common with diffusional processes, parabolic migration in this context refers to migration where the distance traveled is proportional to the square root of the elapsed time t (i.e., L) (Ide 2010).

Kaneko et al. (2018) discovered a continuous seismic signal in a seismogram taken during a tremor burst episode at the shallow part of the Nankai Trough, from the LFE frequency band (1–10 Hz) through the microseismic frequency band (0.1–1 Hz) to the VLFE frequency band (0.01–0.1 Hz), supporting the unified understanding proposed by Ide et al. (2007a). Furthermore, Masuda et al. (2020) discovered that stacked LFE waveforms have a continuous seismic signal from the LFE frequency band (1–10 Hz) to the VLFE frequency band (0.01–0.1 Hz) (Fig. 2b). These researchers stacked numerous seismograms recorded in the Nankai Trough in relation to the timing of high-frequency LFE signals. These findings suggest that very-low-frequency earthquakes and tectonic tremors, or clusters of LFEs, are the high- and low-frequency components, respectively, of a broadband seismic wave originating from the same slow fault slide phenomena. This slow fault slide phenomena has been referred to as a broadband slow earthquake by Kaneko et al. (2018) and Masuda et al. (2020).

The unifying theory put forward by Ide et al. (2007a) regarding geodetic slow earthquakes, however, is still up for discussion (see, for example, Gomberg et al. 2016; Frank and Brodsky 2019; Michel et al. 2019). According to Michel et al. (2019), for instance, seismic moments associated with short-term SSEs in the Cascadia subduction zone are probably going to be proportionate to their cubed lengths (i.e., Mo T3). They contend that the rapid earthquake scaling (the thick red line in Fig. 3a) and the moment-duration scaling of SSEs are comparable. Their theory suggests that distinct moment-duration scaling relations are needed to characterize different types of slow earthquakes because seismic and geodetic slow earthquakes, or even short-term and long-term SSEs, may have different physical sources. Notably, deep short- and long-term SSE measurements in the Nankai Trough are somewhat compatible with a single linear scaling relation, but cannot be explained by a single cubic scaling relation (Ide et al. 2007; Okada et al. 2022).

PREVENTIVE MEASURES

Earthquake Early Warning Risk perception

Introduction

Located in the Tohoku area of Japan on the Pacific coast, Sendai City has a history of frequent earthquakes and tsunamis. In addition to experiencing infrequent, high-risk disasters like the Great East Japan Earthquake of 2011, Sendai is surrounded by active faults. In the meanwhile, King County, Washington State, which encompasses the Seattle metropolitan area, is situated in the Pacific Northwest of the United States on the active faults of the Cascadia Subduction Zone (henceforth referred to as "Seattle"). Both are sizable cities along the coast that are situated in seismically active zones where the locals probably have some experience with earthquakes.

An earthquake early warning system (EEW) can send out a warning a few seconds to a few minutes before intense shaking begins, allowing people who receive it to take preventive measures even though earthquake forecasting is not feasible. In several nations with significant seismic risk, like Japan, Mexico, Romania, Italy, and now the West Coast of the United States, EEW is accessible to the general public. As of May 2021, the U.S. Geological Survey has implemented the ShakeAlert earthquake early warning system (EEWS) across all three West Coast states, alerting both institutional and public users. However, at the time of this research, Washington state had not yet experienced implementation. ShakeAlert is dependent on sensor data from the Advanced National Seismic System; other robust EEWS function as a network of seismic monitoring and alerting systems.

The system's geographically dispersed ground motion sensors identify primary waves, or "p waves," from earthquakes and transmit the data to computer servers for processing. The servers then send out alerts to alert delivery providers, who either notify subscribers directly or trigger automated response systems, such as slowing down trains. Potentially destructive shaking is induced by secondary waves (s waves) from earthquakes. Because s waves travel more slowly than p waves, EEW signals that are broadcast at the speed of light can offer warnings that last anywhere from a few seconds to several minutes. The lead time between the alarm and the onset of potentially harmful shaking at a location increases with the distance between the earthquake's epicenter and that site.

Alerts are broadcast as Wireless Emergency Alert (WEA) messages on radio, television, public sirens, and increasingly smartphones in nations that have an active EEWS, such as Mexico and Japan. The Japanese Meteorological Agency sent out early earthquake warnings (EEWs) to the public, which were quickly shared on TV, radio, and cell phones, just a few seconds before the powerful tremors of the 2011 Great East Japan Earthquake reached the Tohoku area.

According to a poll conducted in Japan, almost 80% of participants who were notified of EEW notifications before the 2011 earthquake were able to take preventative measures.

Currently, Wireless Emergency Alerts and smartphone applications like MyShake and QuakeAlertUSA are used by the EEW alert system ShakeAlert in U.S. West Coast states to notify the public. Many hospitals, transportation networks, and utilities also use alerts to set off automated reactions (such slowing down trains) in the case of severe shaking.

At the time of our study, there was one distinction in the situation between Sendai and Seattle: an EEWS was being developed in Washington state and was not yet widely available, but in Japan, one had been available to the public since 2007. ShakeAlert is an example of a sophisticated EEWS that requires significant infrastructure and technological expenditures to establish and operate successfully. Public support and consensus are essential since covering such large expenses needs public funding, at least in part, if not totally. Simultaneously, the objective of EEWS is to promote protective behavior and enable automatic reactions in situations when severe shaking is anticipated. Therefore, to fully reap the benefits of seismic hazard mitigation, a deeper comprehension of how society and individuals view the system and respond to signals is essential. Furthermore, given the growing effects of catastrophes worldwide, cross-national comparative disaster research has long been seen as crucial and a source of crucial knowledge. These needs are met by the comparative study presented here.

Research aims and questions

In order to enhance comprehension of how people and society view the system and respond to warnings, we provide survey data investigating whether prior conclusions about how people perceive earthquake hazards and EEW are still applicable in the two distinct scenarios of Sendai and Seattle. While both Seattle and Sendai lie close to subduction zones, there are differences in the recent seismic activity and the accessibility of EEW to the general population.

In recent years, Sendai has seen an increased frequency of earthquakes, such as the devastating Great East Japan Earthquake and Tsunami of 2011 with a magnitude of 9, as well as the Miyagi earthquakes of 1978 and 2005 with a magnitude of 7.4 and 7.2, respectively. In contrast, Seattle has had fewer earthquakes in recent years. Since 1965, there have only been five earthquakes of magnitude five or above; the most recent one being the Nisqually Earthquake of 2001, which was magnitude 6.8. The degree to which relationships between earthquake risk perceptions, risk preparedness, and support for governmental policies to reduce earthquake risks can be generalized across these disparate conditions can be examined in light of the similarities and differences in recent events, EEW availability, and seismic risk. In order to do this, we carried out surveys in Seattle and Sendai.

As previously mentioned, we conducted surveys to find out how people in Sendai and Seattle felt about earthquake risks, the usefulness of EEW, and their willingness to pay (WTP) for an EEW smartphone app in Seattle or an upgraded EEW app in Sendai. More precisely, we enquired:

1) How do people in Sendai and Seattle compare in terms of their beliefs, prior experiences, and expectations regarding earthquakes? How well-informed are they

on disaster prevention, and what steps have they taken to be ready for earthquakes? What are the differences between the expected and actual behaviors of inhabitants of Sendai and Seattle after an earthquake and regarding emergency preparedness?

2) Do individuals in both locations think that receiving an earthquake early warning system that might notify them just seconds or minutes before the ground begins to tremble is helpful for personal safety? Does the availability of EEW affect how effective people believe it is in reducing the danger of earthquakes in the two regions?

3) Will consumers be ready to pay for EEW? Are the factors that have been discovered to predict protective action intentions linked to both cities' willingness to pay for EEW?

Risk perception and preventive action intentions predictors

Even though towns in areas like Sendai and Seattle are susceptible to earthquakes, the degree to which individuals perceive danger, as well as their willingness to pay for preventive measures and assess the efficacy of hazard mitigation strategies, vary widely. While some people in high-risk locations are well-prepared for earthquakes, others are not. In a similar vein, certain locals are more ready to spend money on safety precautions than others. People differ in how they receive and understand the information related to risks and uncertainties connected with natural hazards. They also differ in how they perceive risk and in the intents and results of their activity.

Previous studies indicate that behavioral intentions, results, and risk perceptions are explained by a complicated web of factors. It has been demonstrated that exposure to a natural hazard disaster might affect one's sense of the danger associated with that specific hazard.

Heuristics that lessen mental uncertainty can be derived from experience. The affect heuristic is one such heuristic that uses emotional reactions in place of cognitive risk evaluation.

Unfavorable feelings like dread or terror likely to contribute to increased perceived risks. These emotions can be utilized to measure the risk linked with earthquakes later on.

People may also rely on past experiences or a value that was first offered when assessing risk by using availability or anchoring heuristics [19]. As a result, people's perceptions of future dangers are probably influenced by the severity and frequency of natural hazard encounters, such as earthquakes, as well as the feelings connected to them.

Experience and perception of risk are two important indicators of readiness. Predictors like the characteristics of the risk itself and the resources needed to get ready for it have frequently shown strong correlations in earlier research.

In addition, opinions on the degree to which a particular hazard mitigation strategy provides protection against a natural disaster differ across people and are typically linked to readiness. Perceptions of risk and the efficacy of risk mitigation strategies, such as early earthquake warning systems, have been linked not just to preparation but also to the willingness to pay for them.

According to earlier survey-based research, the majority of Japanese people believe that EEW is beneficial.

These studies highlight how an EEW system cannot successfully improve public safety during an earthquake unless it is combined with public education and training. According to a prior survey-based research carried out in the state of Washington, the majority of respondents thought that EEW would be beneficial for their personal safety in the event of an earthquake. According to this study, participants who had already experienced an earthquake believed they were more likely to suffer personal injury from one than those who had not.

One method that is frequently used to determine the value of a non market commodity is willingness to pay (WTP) estimation. WTP has been used, for instance, to evaluate the worth of additional danger early warning systems. Usually, choice experiments or contingent valuation are used to estimate it. The responses to the WTP question in this study were elicited as—and are interpreted as—a relative expression of interest and support in EEW, rather than as an accurate or reliable measure of how much someone might pay for EEW, given the known susceptibility of stated WTP responses to various framing effects (e.g., anchoring).